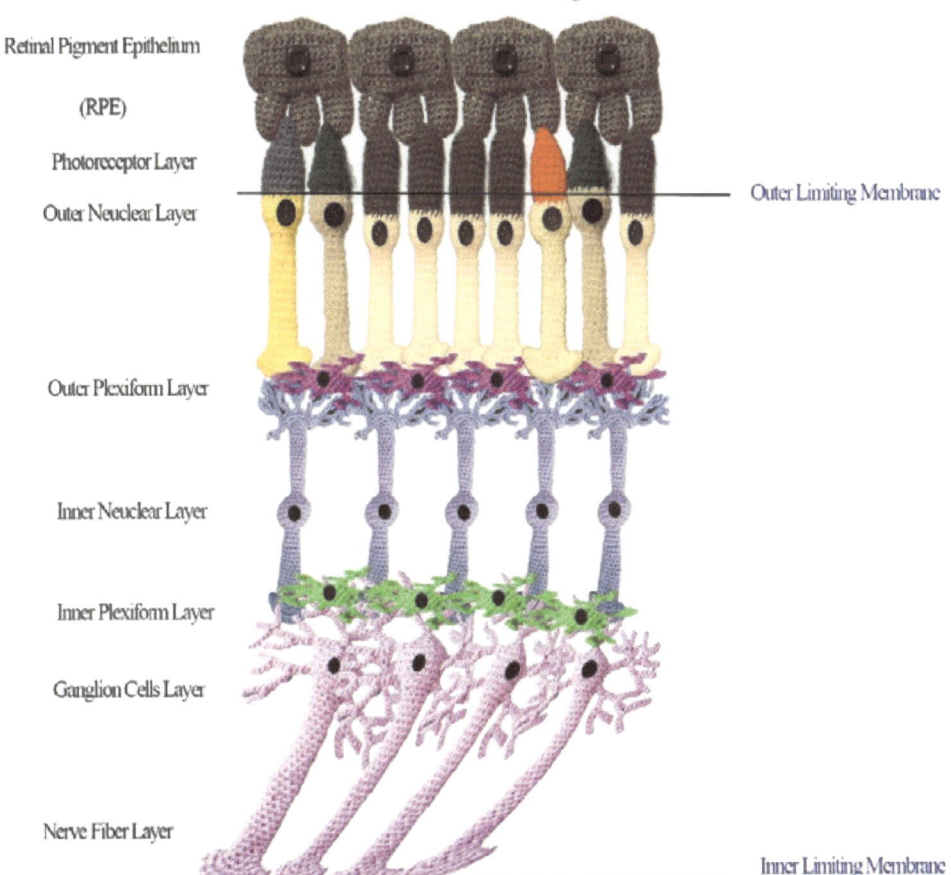

© 2019 The Purple Lilac Science Crochet. Tahani Baakdhah- All rights reserved. These patterns are FOR PERSONAL USE ONLY! The pattern or parts of it may not be reproduced, published (online or printed), altered or resold. You can sell items made from this pattern, provided that they are handmade by yourself in a limited number and you give credit to the designer. Please add the following lines to your item description and any item you are selling: This pattern is handmade by (your name) from a design and pattern by Tahani Baakdhah (@thepurplelilac).

TABLE OF CONTENTS

Who I am 4
Abbreviations 5
Special techniques 6
    Slipknot 6
    Chain 2 6
    Magic Ring 7
    Single Crochet (SC) 8
    Increase 9
    Decrease 9
The Retina 10
Retinal Stem Cell Sphere Pattern 11
Cone Photoreceptor Pattern 13
Rod Photoreceptor Pattern 18
Retinal Pigment Epithelium Pattern 22
Bipolar Cell Pattern 25
Ganglion Cell Pattern 28
Horizontal Cell/ Amacrine Cell Pattern 31
Eye Pattern 33

WHO I AM

My name is Tahani Baakdhah, MD, MSc and PhD candidate at the University of Toronto. I study stem cells in the eye, how they develop and make the retina, how different cells within the retina work and how we can replace them in case of disease or injury. I believe that art is a powerful tool to educate, illustrate and translate complex science concepts. With my crochet skills and artistic soul and mind, I turned my PhD project into art. With hook and yarn I build the retina, not with stem cells, but with stitches, one stitch at a time. Using my creations as models to communicate my research with the public, students and users on social media. My passion and creativity are the fuel and my skills are my tools to create, inspire and motivate others.

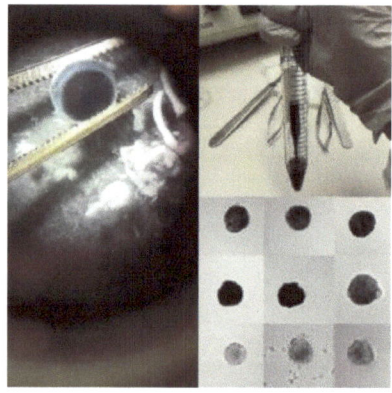 

  *@thepurplelilac*

ABBREVIATIONS

ch	chain
SC	single crochet
Sl st(s)	slip stitch (es)
Inc	increase
dec	decrease
R	round

SPECIAL TECHNIQUES

Slipknot

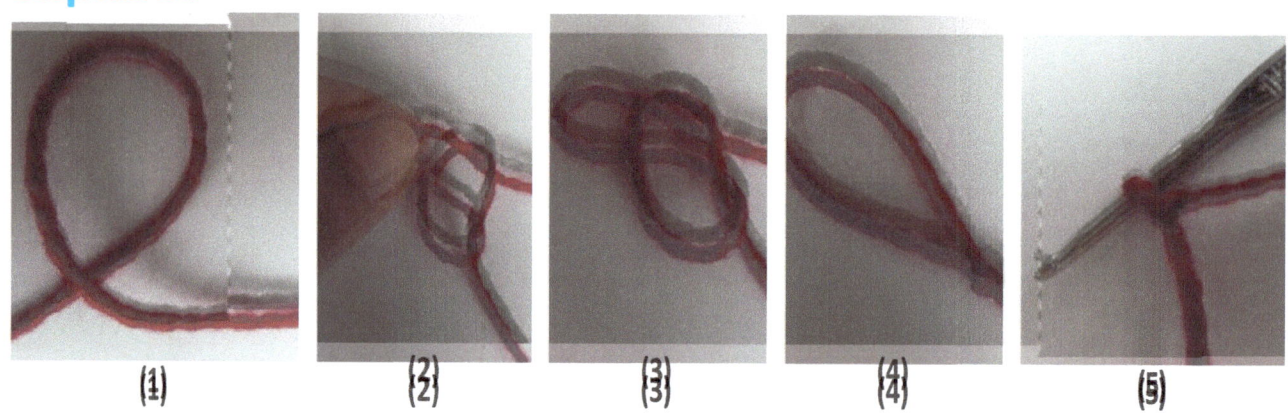

(1)　(2)　(3)　(4)　(5)

Chain 2

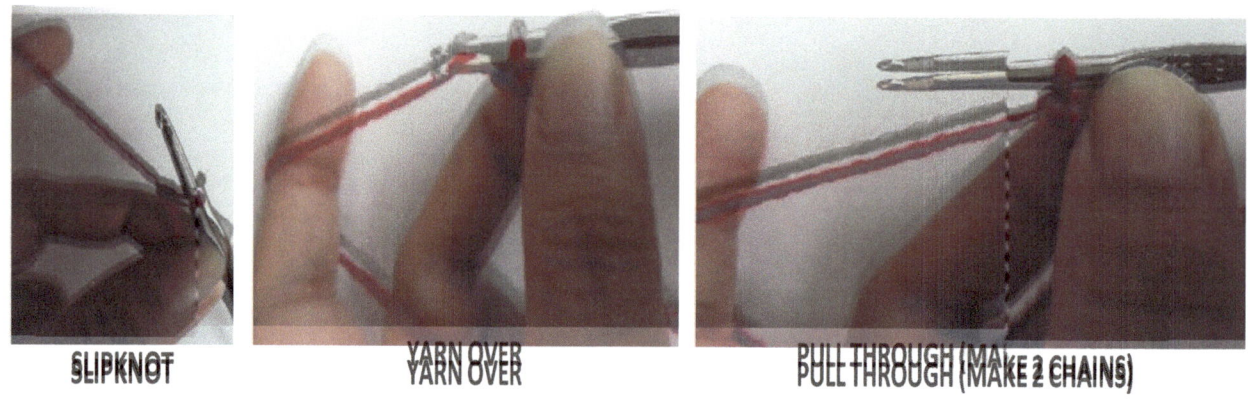

SLIPKNOT　　YARN OVER　　PULL THROUGH (MAKE 2 CHAINS)

Magic Ring

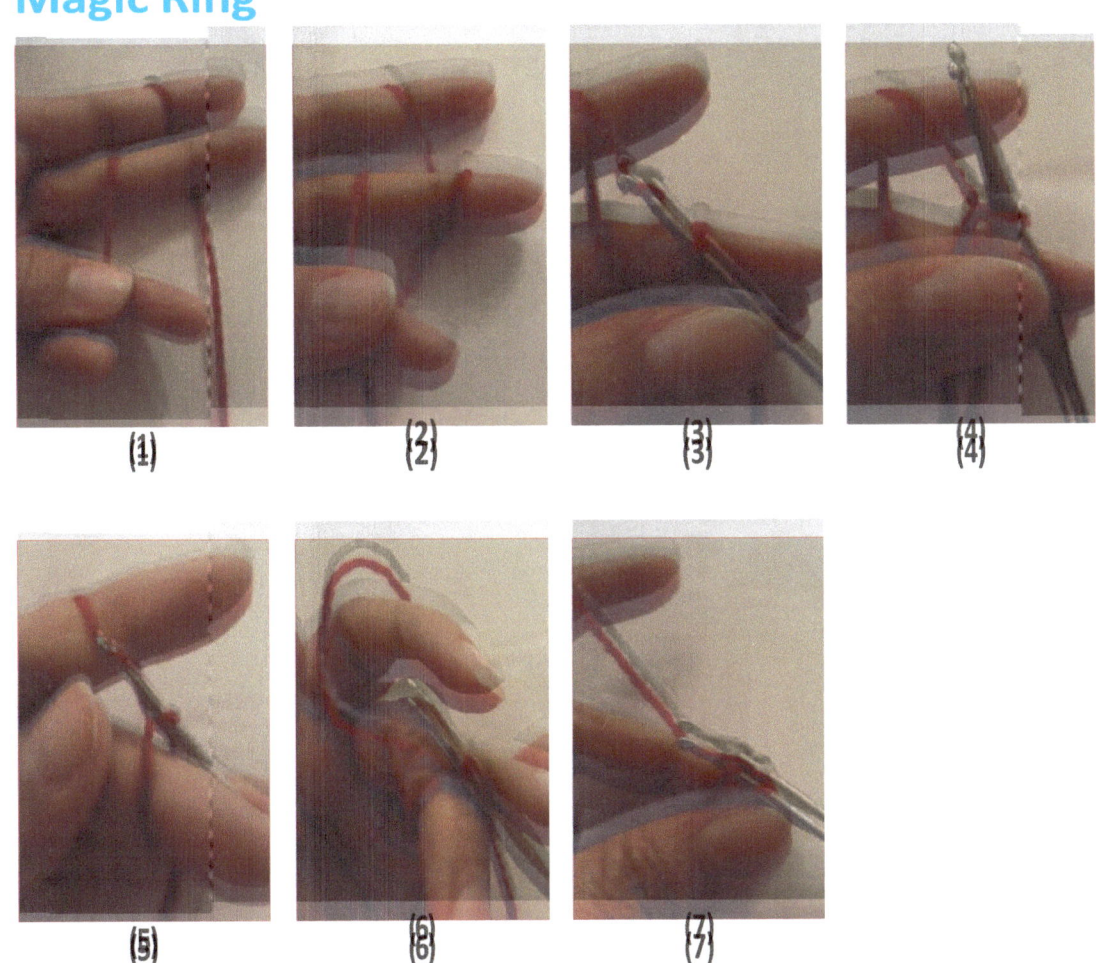

(1) (2) (3) (4)
(5) (6) (7)

Single Crochet (SC)

(1) INSERT YOUR HOOK INTO THE SECOND CHAIN FROM THE HOOK

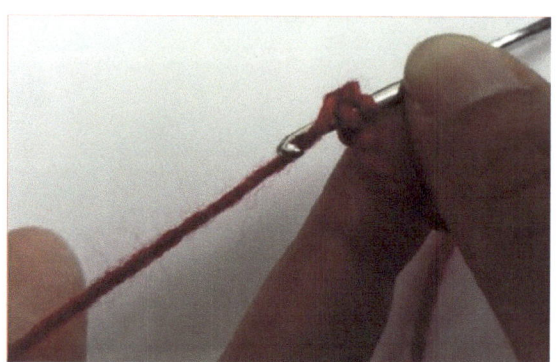

(2) YARN OVER

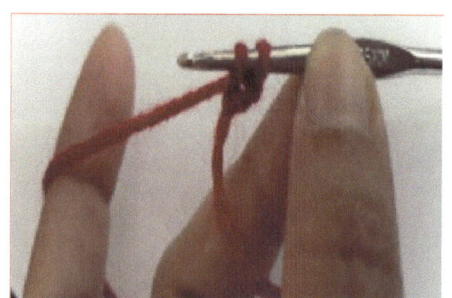

(3) PULL THROUGH (NOW YOU HAVE TWO LOOPS ON THE HOOK)

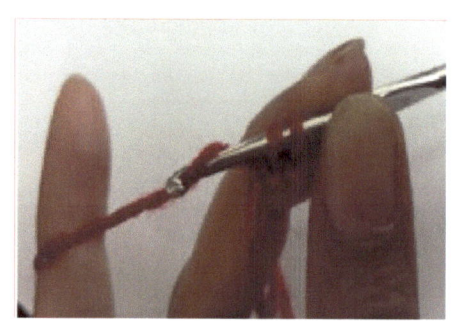

(4) YARN OVER

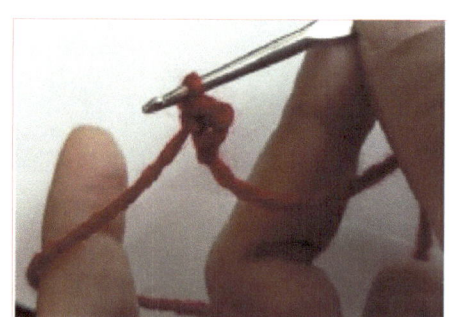

(5) PULL THROUGH (NOW YOU HAVE ONE LOOP ON THE HOOK)

Increase

ROW (1)

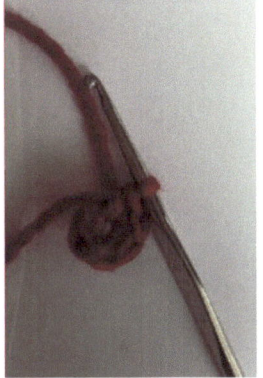

6 (SC) IN A MAGIC RING (TOTAL= 6 SC)

ROW (2)

2 (SC) IN EACH (SC) FROM THE PREVIOUS ROW (TOTAL=12 SC)

Decrease

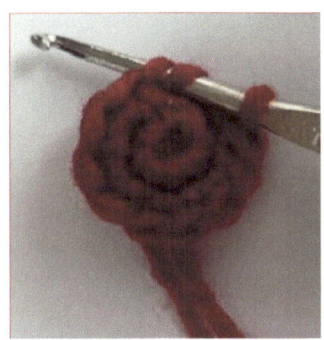

(1) INSERT YOUR HOOK INTO THE FRONT LOOP OF THE NEXT 2 STITCHES

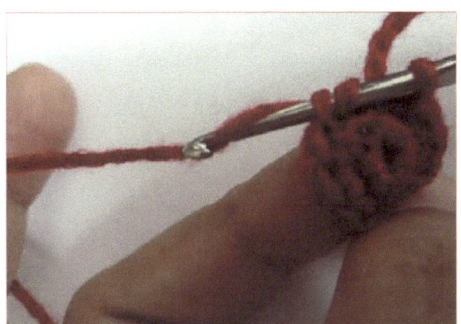

(2) YARN OVER

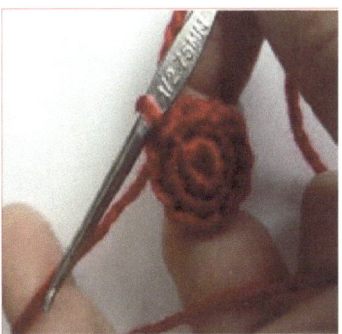

(3) PULL THROUGH

THE RETINA

The retina is the inner-most layer of your eye that converts the light energy to electrical signals through specialized light sensing neurons. These signals are carried to the brain by the optic nerve to interpret the images we see.

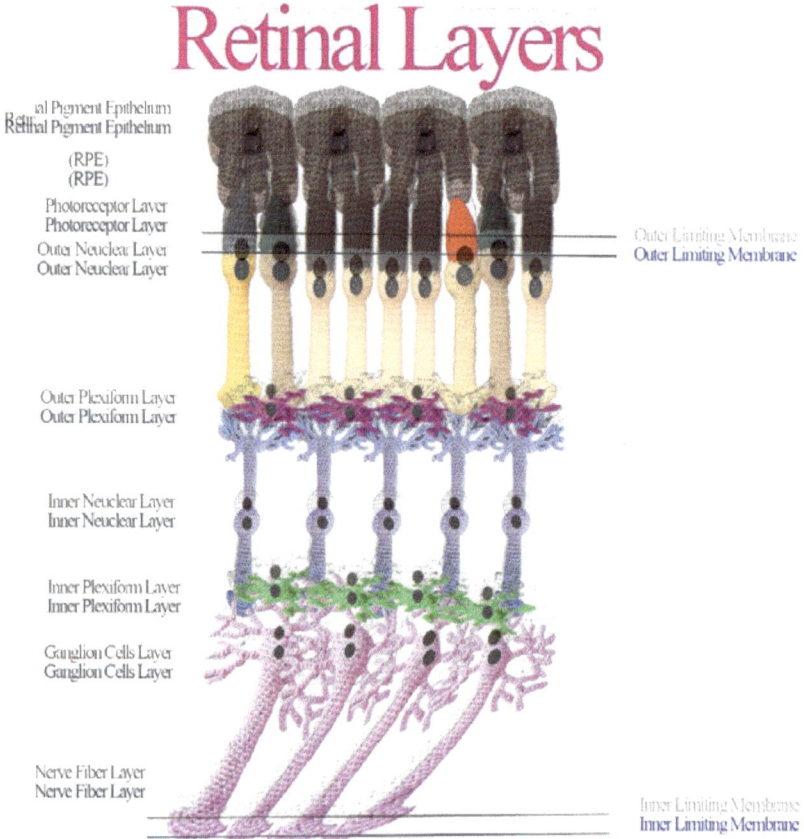

RETINAL STEM CELL SPHERE PATTERN

Use white yarn (light worsted size 3 yarn). Hook size: 3.5 mm.

R1: Start with making 6 SC in a magic ring or chain 2 (6).

R2: 2 SC in every stitch around (12).

R3: "1 SC in the first stitch then 2 SC in the next stitch" repeat to the end (18).

R4: "2 SC then 2 SC in the next stitch" repeat to the end (24).

R5: "3 SC then 2 SC in the next stitch" repeat to the end (30).

R6-9: SC around (30).

Start stuffing from now on.

R10: "3 SC then decrease" repeat to the end (24).

R11: "2 SC then decrease" repeat to the end (18).

R12: "1 SC then decrease" repeat to the end (12).

R13: decrease all around (6).

R14: decrease all around (0).

Fasten off and leave long tail for sewing.

Using glue gun stick small pompoms (black, white and gray) on the crocheted ball in a random arrangement (be creative).

RETINAL STEM CELLS (RSCS)

Scientists grow RSCs in the lab inside incubators and differentiate them to all retinal cell types.

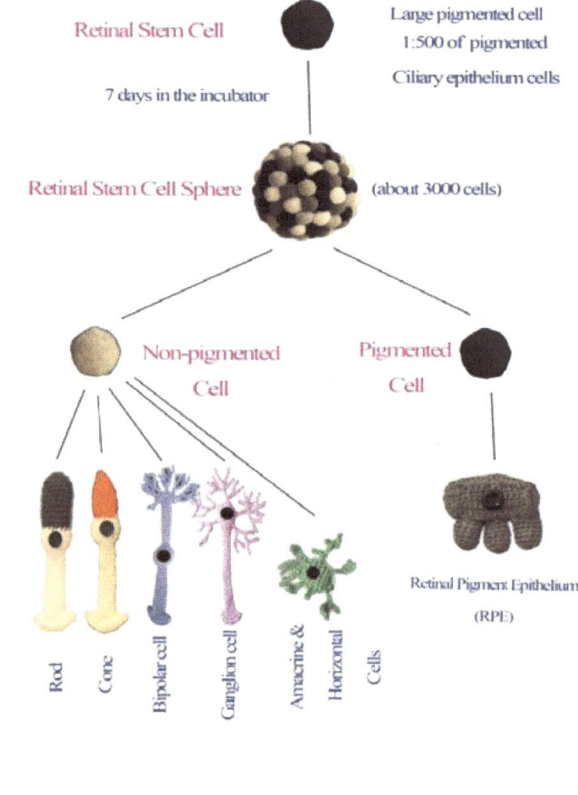

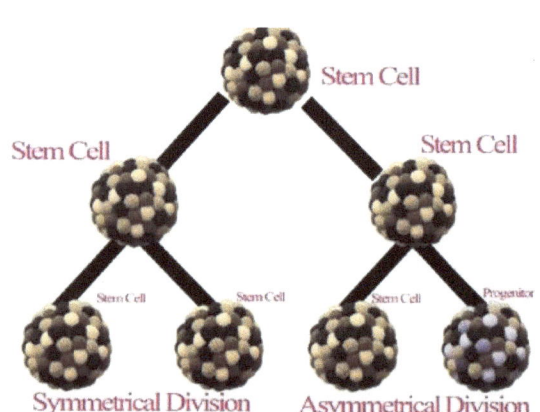

Are large pigmented cells found at the edge of the retina in an area called the "Ciliary Epithelium". During development they divide symmetrically giving two daughter stem cells or asymmetrically giving one stem cell and one progenitor cell (can give any specialized retinal cell types).

CONE PHOTORECEPTOR PATTERN

Use light worsted size 3 yarn. Hook size: 3.5 mm.

Cone (make 3, one using red yarn, another using green yarn and a third using blue yarn).

R1: Start by making 6 SC in a magic ring or chain 2 (6).

R2: "1 SC in the first stitch then 2 SC in the next stitch" repeat to the end (9).

R3: "2 SC followed by 2 SC in the next stitch" repeat to the end (12).

R4-9: SC around (12).

R10: 5 SC, 3SC in the same stitch, 5 SC followed by 3 SC in the last stitch (16).

R11: SC all around (16).

Fasten off and leave long tail for sewing.

Proceed to the cell body pattern.

CONE PHOTORECEPTOR PATTERN

Cell Body (use white yarn)

R1: 6 SC in a magic ring or chain 2 (6).

R2: 2 SC in each stitch around (12).

R3: "1 SC in the first stitch followed by 2 SC in the next stitch" repeat to the end (18).

R4: "2 SC followed by 2 SC in the next stitch" repeat to the end (24).

R5-7: SC all around (24).

R8: "2 SC followed by decrease" repeat to the end (18). and start stuffing the body.

R9: "1 SC followed by decrease" repeat to the end (12).

R10: decrease all around (6).

Stuff the body. Don't fasten off.

Proceed to the axon pattern.

CONE PHOTORECEPTOR PATTERN

Axon
R11-22 (or to the desired length): SC around (6).
Don't fasten off. Proceed to the axon terminal pattern.

Axon Terminal
1: 2 SC in each stitch from the previous row (12).
2: 2 SC in each stitch around (24).
3: Fasten off and leave long tail to sew the two ends.

Installing parts together
Sew cone to cell body then sew a button to the center of the cell body to make the nucleus.

CONE PHOTORECEPTORS

Blue Cone   Green Cone   Red Cone

Are photosensitive cells in the retina responsible for color vision. There are three types of light-sensitive cone cells in the retina of the eye, each type is sensitive to a different wavelength: Blue cone is sensitive to short wavelengths (S-cone), Green cone is sensitive to medium wavelengths (M-cone) and Red cone is sensitive to long wavelengths (L-cone). There are about 6 million cones in each retina concentrated mainly at the center of the retina in an area called "Fovea".

CONE PHOTORECEPTORS

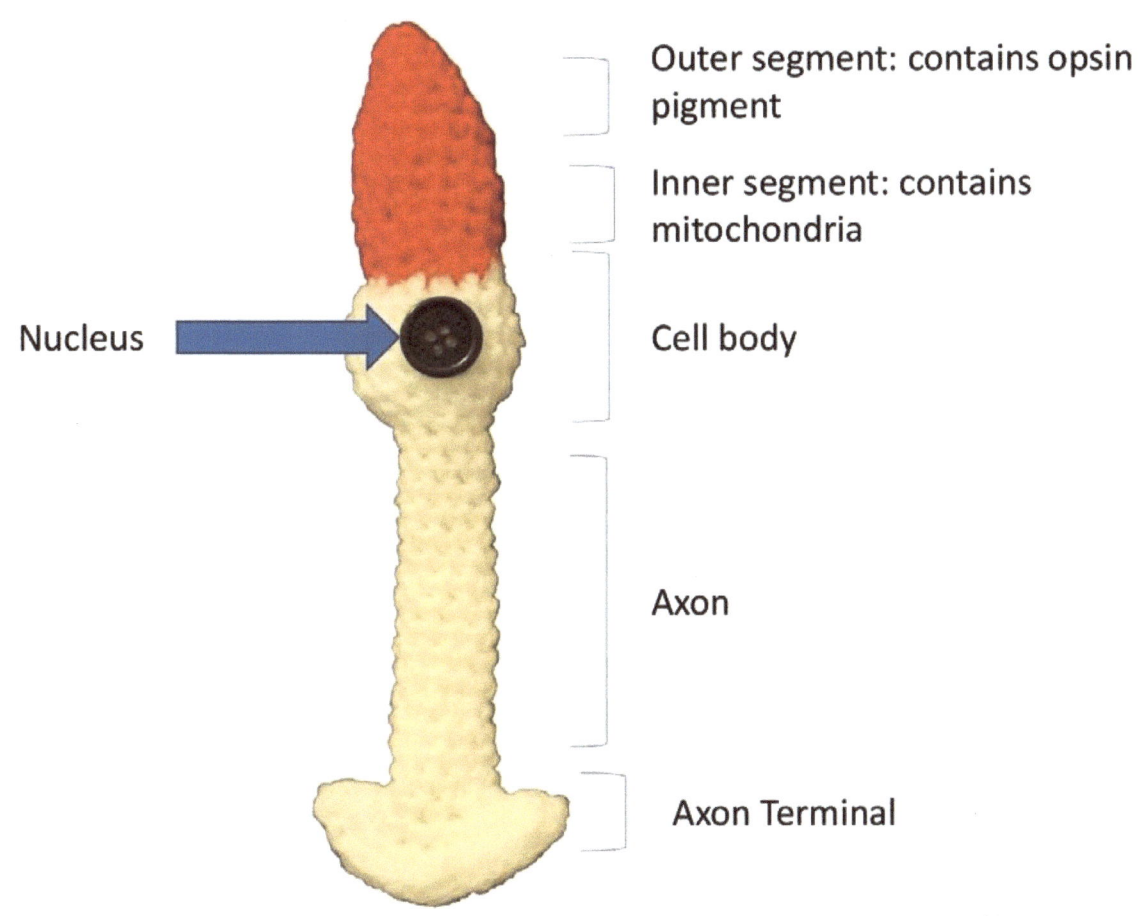

Outer segment: contains opsin pigment

Inner segment: contains mitochondria

Nucleus

Cell body

Axon

Axon Terminal

ROD PHOTORECEPTOR PATTERN

Use gray yarn (light worsted size 3 yarn). Hook size: 3.5 mm

Rod

R1: Start with making 6 SC in a magic ring or chain 2 (6).

R2: 2 SC in every stitch around (12).

R3: "1 SC in the first stitch then 2 SC in the next stitch" repeat to the end (18).

R4-10: SC around (18).

Stuff the rod. Fasten off and leave long tail for sewing. Proceed to the cell body pattern.

ROD PHOTORECEPTOR PATTERN

Cell Body (use white yarn)

R1: 6 SC in a magic ring or chain 2 (6).

R2: 2 SC in each stitch around (12).

R3: "1 SC followed by 2 SC in the next stitch" repeat to the end (18).

R4: "2 SC followed by 2 SC in the next stitch" repeat to the end (24).

R5-7: SC all around (24).

R8: "2 SC followed by decrease" repeat to the end (18). and start stuffing the body.

R9: "1 SC followed by decrease" repeat to the end (12).

R10: decrease all around (6).

Stuff the body. Don't fasten off.

Proceed to the axon pattern.

ROD PHOTORECEPTOR PATTERN

Axon

R11- 22 or to the desired length: SC around (6).

Don't fasten off. Proceed to the axon terminal pattern.

Axon Terminal

1. 2 SC in each stitch from the previous row (12).
2. 2 SC in each stitch around (24).
3. Fasten off and leave long tail to sew the two ends.

Installing parts together

Sew rod to cell body then sew a button to the center of the cell body to make the nucleus.

ROD PHOTORECEPTORS

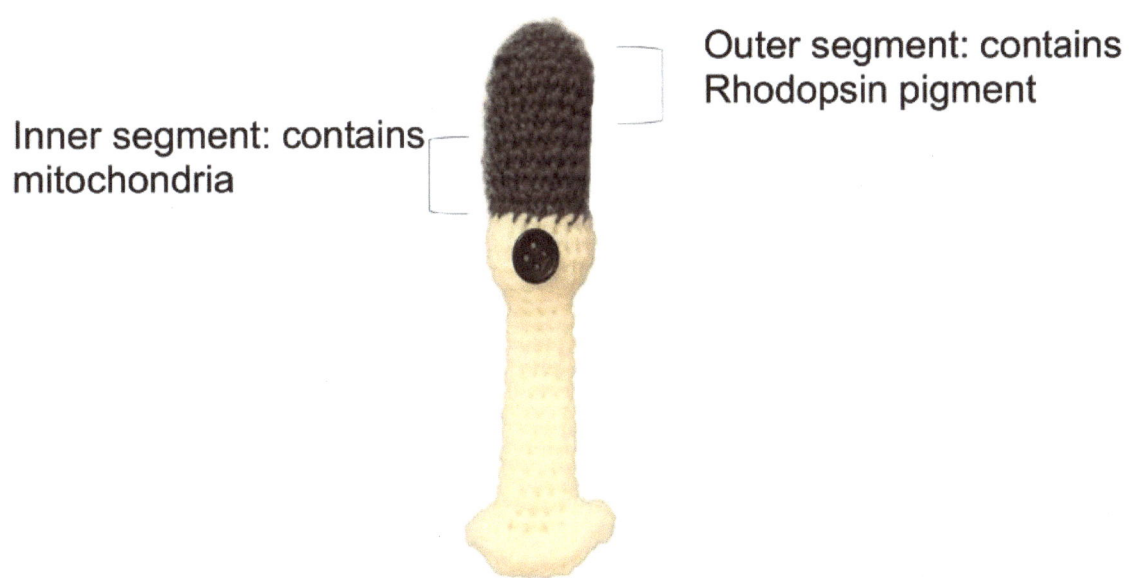

Inner segment: contains mitochondria

Outer segment: contains Rhodopsin pigment

Are so sensitive to low light and enable us to see in the dark. There are about 12 million rods in one human eye concentrated more at retinal periphery. Their outer segment contains a light sensitive pigment called Rhodopsin. This pigment is derived from vitamin A and deficiency of this vitamin can lead to a condition called "Night-Blindness".

RETINAL PIGMENT EPITHELIUM PATTERN

Use gray yarn (light worsted size 3 yarn).

Hook size: 3.5 mm.

Cell Body (part1)

R1: chain 15 (15).

R2-8: Turn, chain 1 then 1 SC in each stitch to the end (15).

R9: 1 SC in the front loop only to the end.

R10: SC in the back loop only to the end.

R11-18: 1 SC in every stitch to the end.

Fill in using polyester filling (you will have a rectangle formed by the end of this stage).

Fasten off and proceed to the part 2 pattern.

Sew black button to the center.

RETINAL PIGMENT EPITHELIUM PATTERN

Cell Body (part2)

R1: chain 15 (15):

R2-8: Turn, chain 1 then 1 SC in each stitch to the end (15):

Fasten off, leave long tail for sewing to join part 1 and part 2 together.

Cell Process (villi) Make 3

R1: 6 SC in a magic ring or chain 2 (6):

R2: 2 SC in each stitch around (12):

R3: "1 SC in the first stitch followed by 2 SC in the next stitch" repeat to the end (18):

R4-9: SC all around (18):

Fill with polyester filling, fasten off, leave long tail for sewing to part 2:

RETINAL PIGMENT EPITHELIUM

The inner most pigmented cell layer of the retina. They absorb extra waves and light to protect the eye and prevent retinal damage. They also support photoreceptors and keep them healthy and functioning.

BIPOLAR CELL PATTERN

Use light worsted size 3 yarn. Hook size: 3.5 mm
Color: any of your choice.

Cell Body

R1: chain 6 then slip stitch in the first chain (6).

R2: 2 SC in each stitch around (12).

R3: "1 SC in the first stitch followed by 2 SC in the next stitch" repeat to the end (18).

R4: "2 SC followed by 2 SC in the next stitch" repeat to the end (24

R5-7: SC all around (24).

Start stuffing the body.

R8: "2 SC followed by decrease" repeat to the end (18) and start stuffing the body.

R9: "1 SC followed by decrease" repeat to the end (12).

R10: decrease all around (6).

Stuff the body. Don't fasten off.

Proceed to the axon pattern.

BIPOLAR CELL PATTERN

Axon (make 2 on the top and lower holes)
R11- 22 (or to the desired length: SC around (6).
Don't fasten off. Proceed to the axon terminal pattern.

Axon Terminal (1)
1. 2 SC in each stitch from the previous row (12).
2. 2 SC in each stitch around (24).
3. Fasten off and leave long tail to saw the two ends.

Axon Terminal (2)
Start anywhere on the axon terminal, attach your yarn by slip stitch.
1. Chain 9 then 4 slip stitch in every stitch back the way toward the body. Chain 4 and slip stitch all the way to the cell body.
2. Skip 3 stitches and repeat Step 1.
3. Continue till you reach the last stitch on the body.

Fasten off and wave the tail.
You can add more extensions with different length (use your imagination).

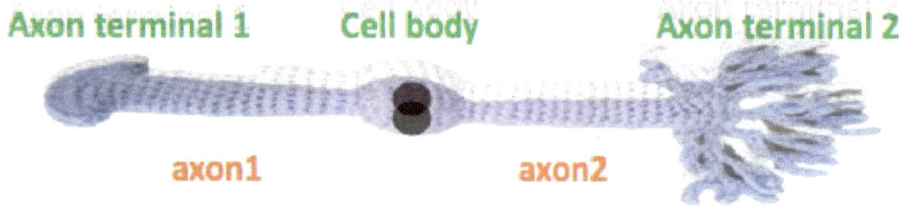

Axon terminal 1 — Cell body — Axon terminal 2
axon1 — axon2

BIPOLAR CELLS

Photoreceptor Layer

Bipolar cell, as the name indicates, have two axons arising from the cell body. One connects it to a photoreceptor and the other connect it to a ganglion cell.

Ganglion cell Layer

GANGLION CELL PATTERN

Use yarn color of your choice
(light worsted size 3 yarn and 3.5mm hook).

Cell Body

R1: Start with making 6 SC in a magic ring or in chain 2 (6).

R2: 2 SC in every stitch around (12).

R3: "1 SC in the first stitch then 2 SC in the next stitch"
repeat to the end (18).

R4: "2 SC then 2 SC in the next stitch"
repeat to the end (24).

R5-6: SC around (24).

Start filling in using polyester filling.

R7: "2 SC followed by decrease"

repeat to the end (18).

R8: "1 SC followed by decrease" repeat to the end (12).

R9: decrease all around (6).

Don't fasten off. Proceed to the axon pattern.

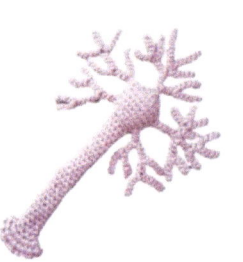

GANGLION CELL PATTERN

Axon

R10- 22 or until you reach the desired length: SC around (6).

Don't fasten off. Proceed to the axon terminal pattern.

Axon Terminal

1. 2 SC in each stitch from the previous row (12).
2. 2 SC in each stitch around (24).
3. Fasten off and leave long tail to sew the two ends.

Dendrites

Start anywhere on the cell body, attach your yarn by slip stitch.

1. Chain 9 then 4 slip stitch in every stitch back the way toward the body. Chain 4 and slip stitch all the way to the cell body.
2. Skip 3 stitches and repeat Step 1.
3. Continue till you reach the last stitch on the body.

Fasten off and wave the tail.

You can add more extensions with different length (use your imagination).

GANGLION CELLS

Electrical signals are transmitted from the photoreceptors to our brain's visual cortex which is located in the occipital lobe through ganglion cell fibers that come together to form the optic nerve.

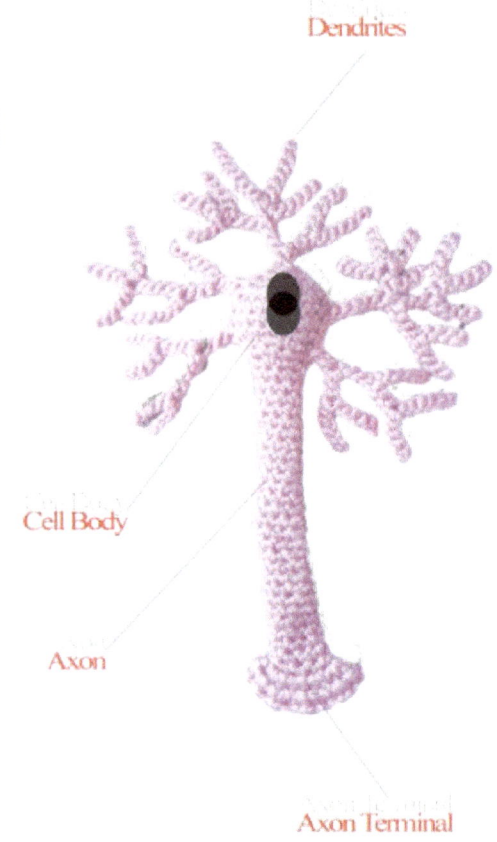

HORIZONTAL CELL/ AMACRINE CELL PATTERN

Use any yarn color you prefer (light worsted size 3 yarn and 3.5 mm hook):

Cell Body

R1: 6 SC in a magic ring or chain 2 (6).

R2: 2 SC in each stitch around (12).

R3: "1 SC in the first stitch followed by 2 SC in the next stitch" repeat to the end (18).

R4: "2 SC followed by 2 SC in the next stitch" repeat to the end (24).

R5-9: SC all around (24).

R10: "2 SC followed by decrease" repeat to the end (18) and start stuffing the body.

R11: "1 SC followed by decrease" repeat to the end (12).

R12-13: decrease all around (6).

Don't fasten off and proceed to the next steps to make the dendrites.

Dendrites (1)

Chain 14, Sl st back 5 stitches, chain 5, Sl st 6 stitch back, chain 6, Sl st 6 stitches back then continue Sl st in 3 stitches.

Chain 6, Sl st 6 stitches back, Chain 6 then Sl st 6 stitches back.

Continue 4 Sl st down the way.

Chain 8, Sl st 5 chain back, chain 6, Sl st 5 stitches back.

Sl st 2 stitches then continue on the main stem Sl st 4 stiches then fasten off.

Dendrites (2)

Attach your yarn to the opposite pole by Sl st then do the same steps as Dendrites (1).

HORIZONTAL CELLS/AMACRINE CELLS

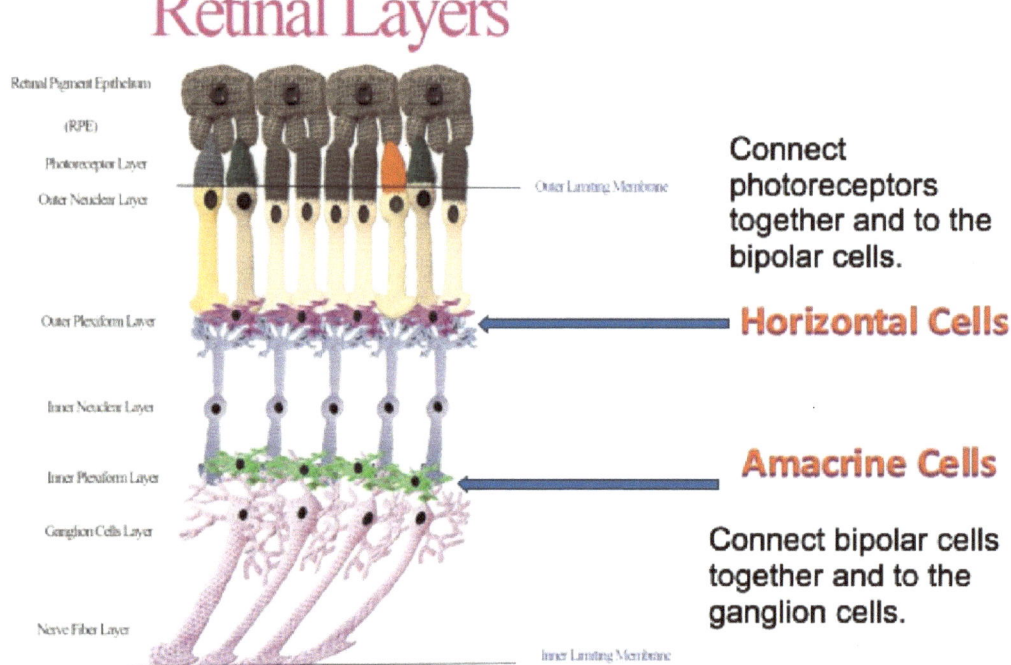

EYE PATTERN

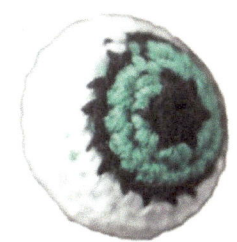

Use light worsted size 3 yarn and 3.5mm hook.

Pupil

R1: Using black yarn start by making 6 SC in a magic ring or chain 2.

Join any other color you like and proceed to make the iris.

Iris

R2: 2 SC in every stitch around (12).

R3: SC around (12).

Change to black yarn join and proceeds to the Ciliary Epithelium pattern.

Ciliary Epithelium

R4: SL St around (12).

Change to white yarn, join and proceed to the next row.

EYE PATTERN

Sclera

R5: working in the front loop only crochet the following:
"1 SC in the first stitch then 2 SC in the next stitch" repeat to the end (18).

R6: working in the back loop only
"2 SC then 2 SC in the next stitch" repeat to the end (24).

R7-8: SC around (24).

R9: "2 SC followed by decrease" repeat to the end (18) and start stuffing.

R10: "1 SC followed by decrease" repeat to the end (12).

R11-13: decrease all around.

Fasten off.

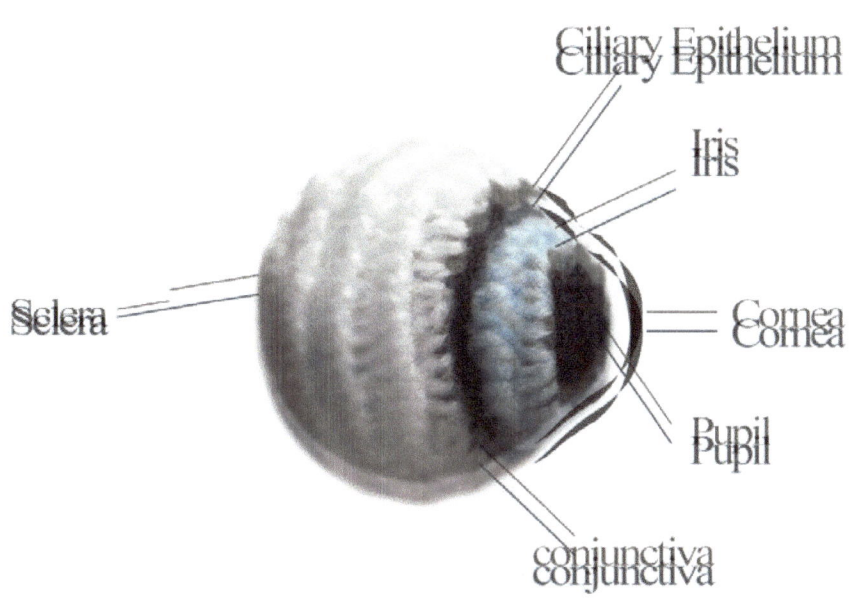

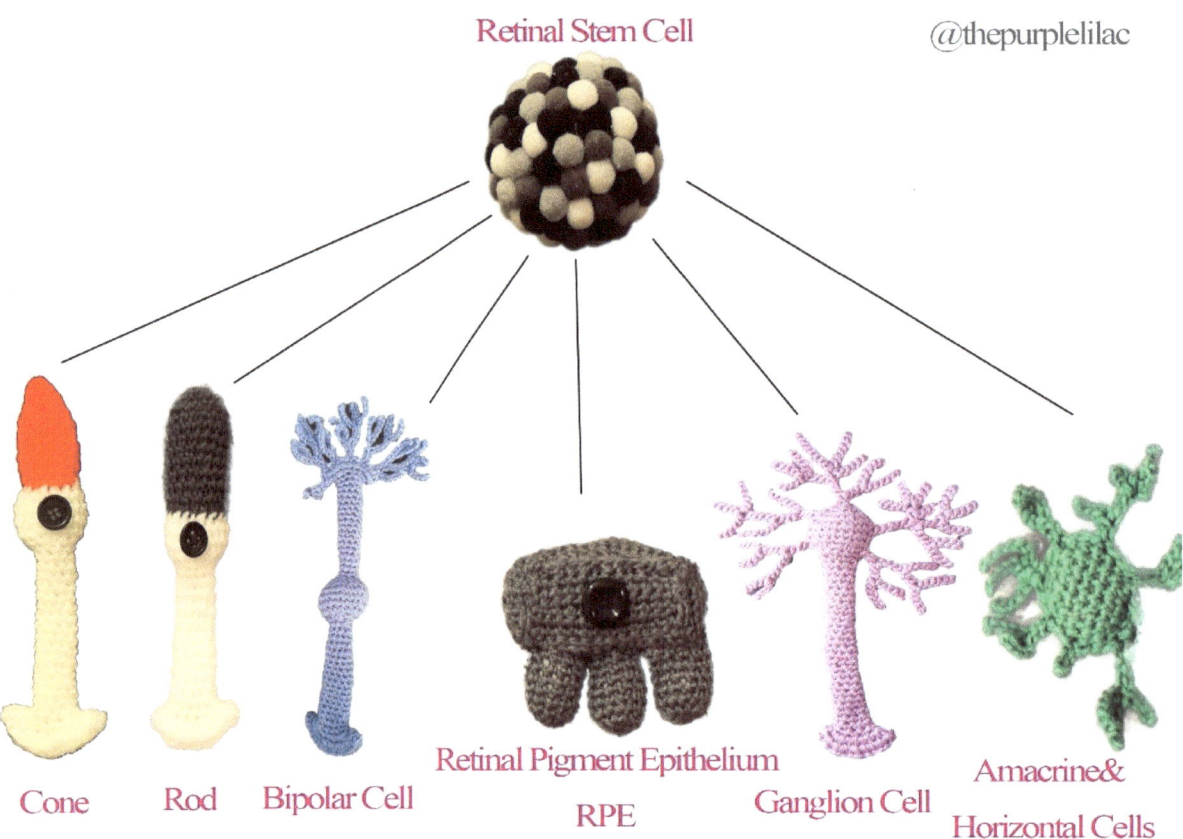

www.ingramcontent.com/pod-product-compliance
Lightning Source LLC
Chambersburg PA
CBHW051936210526
45473CB00006B/2269